Katydids
The Singing Insects

Books by Barbara Ford

KATYDIDS: THE SINGING INSECTS
HOW BIRDS LEARN TO SING
CAN INVERTEBRATES LEARN?

Katydids
The Singing Insects

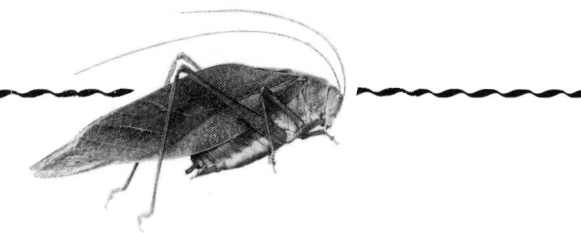

by Barbara Ford

Illustrated with photographs

drawings by William Jaber

JULIAN MESSNER   New York

Published by Julian Messner, a Division of Simon & Schuster, Inc.
A Gulf + Western Company
1 West 39 Street, New York, N.Y. 10018. All rights reserved.

Copyright © 1976 by Barbara Ford

Printed in the United States of America

Design by Ruth Bornschlegel

Library of Congress Cataloging in Publication Data
Ford, Barbara.
  Katydids: the singing insects.

  Includes index.
  SUMMARY: Describes the physical characteristics and life cycle of various members of the katydid family.
  1. Katydids—Juvenile literature. [1. Katydids. 2. Insects] I. Jaber, William. II. Title.
QL508.T4F67   595.7'26   76-29059
ISBN 0-671-32814-X

To Doug, who drove thousands of miles in search of katydid experts

Photo Credits

The Church of Jesus Christ of Latter-day Saints: pp. 51, 52
Jerome Freilich: pp. 62, 63
Dr. Glenn K. Morris: pp. 31, 40, 45, 46, 47
Nevada Department of Agriculture: pp. 50, 54
Dr. David C. Rentz: pp. 22, 53, 57, 58, 59
Charlie E. Rogers: p. 71
U.S. Department of Agriculture: Title page, pp. 25, 26, 68, 73
Chuck Woods: pp. 20, 38

Acknowledgment

My special thanks to Dr. Glenn K. Morris of the University of Toronto, Dr. David C. Rentz of the California Academy of Sciences, and Dr. Thomas J. Walker of the University of Florida for their help in preparing this book. Without their aid, there would be no book. Dr. Stanley K. Gangwere of Wayne State University, Dr. Charlie Rogers and Charles R. Parencia of the U.S. Department of Agriculture, F. Charles Graves of the Church of Jesus Christ of Latter-Day Saints, Harry Gallaway of the Nevada State Department of Agriculture and Dr. D. K. McE. Kevan of McGill University also furnished valuable assistance.

Contents

1. The Secret Singers · 11
2. A Katydid's Life · 21
3. The Song of the Katydid · 29
4. A Katydid Laboratory · 40
5. Katydid Pests · 49
6. A Katydid Collector · 57
7. The Katydid's Relatives · 66
8. How to Find, Catch, and Keep Katydids · 75
Biological Supply Houses · 88
Glossary · 89
Index · 93

1

The Secret Singers

"Ka ki–KAK! Ka ki–KAK!"

Have you ever heard this sound on a summer night?

Many people in the United States have. It is the song of the "true katydid," the best known of the many kinds of katydids. Green and a little more than an inch long, it has large wings and long, thin *antennae*—the threadlike "feelers" on the front of its head. The song of the true katydid sounds like the words "katy did", at least to some people, so this insect and all its relatives are called "katydids."

There are many different kinds of katydids, some of them living in all parts of the United States, but people seldom see them. You are much more likely to *hear* a katydid than *see* it. Katydids are secretive animals. Each one lives by itself, not in a group, and avoids humans. And many of them are active mainly at night, when we cannot see them.

Some katydids are active during the day, but even those are hard to find because of their protective coloring. Katydids blend in with their surroundings. The wings of many katydids are leaf green, and shaped something like leaves. They even have veins that look just like the veins in a leaf. The wings may even have what looks like little holes and torn places, just like those in leaves.

When a katydid like this sits in a bush or tree, with its wings pointing in the same direction as the leaves, it is very hard to see.

Why does a katydid try to look like a leaf?

To protect itself. Katydids do not move very quickly, and they make an appetizing dinner for some birds and other animals. To avoid capture, katydids make themselves look like something else—a leaf. Scientists call animals like this "leaf mimics." A mimic is an

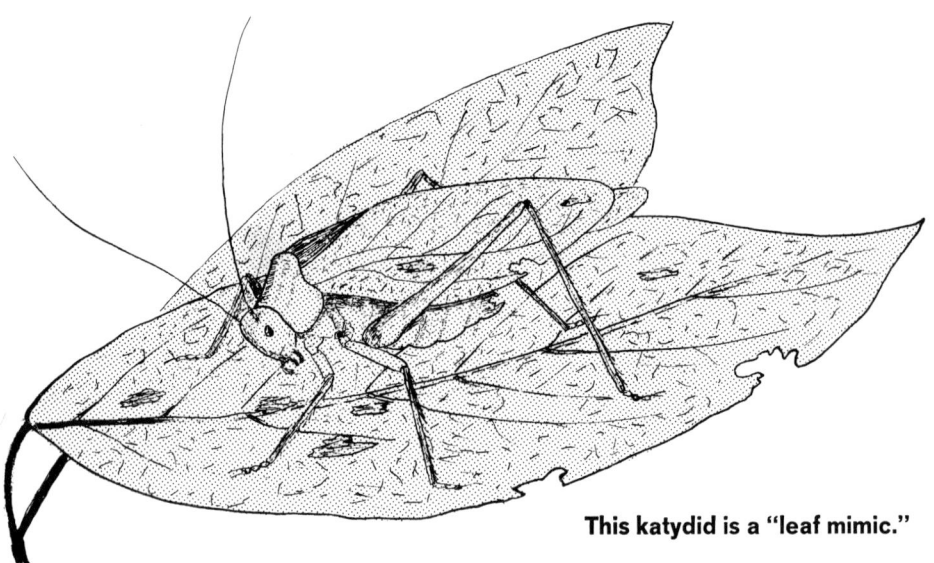

This katydid is a "leaf mimic."

animal that looks, sounds, or acts like another animal or thing.

The katydid is sometimes confused with the cricket or shorthorned grasshopper. All three are related, but they belong to different "families." In science, the word family has a special meaning.

Many years ago, a scientist named Linnaeus worked out a system of grouping animals and plants so that they could be studied more easily. Under his system, which now includes some modern additions, katydids belong to a large and varied group of animals known as the insects. All insects have six legs and a body with three parts: head, chest or thorax, and abdomen. The insects are divided into smaller groups called "orders." The order to which katydids belong is *Orthoptera*. Orthoptera is a Latin word that means "straight-winged." Most orthopterans have straight wings.

The orthopterans, in turn, are divided into families, which are groups with many characteristics, or features, in common. Most of these families we know: the cockroaches, the walkingsticks, the praying mantis, the crickets, and the grasshoppers. There are two grasshopper families, the shorthorned grasshoppers and the longhorned grasshoppers. The words "shorthorn" and "longhorn" refer to the length of the antennae. Katydids belong to the longhorned grasshopper family because they have long, slender anten-

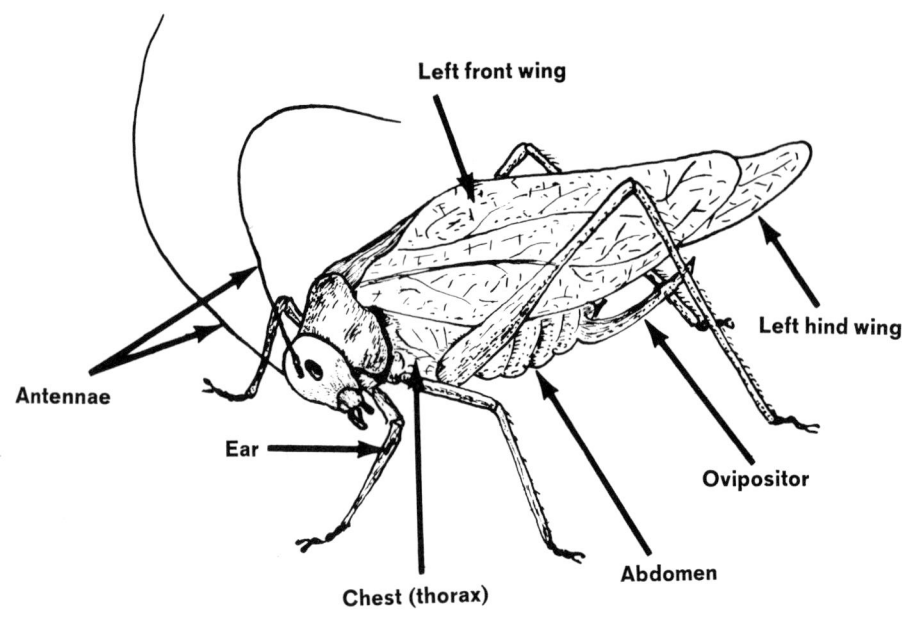

A three-quarter view of an angular-winged katydid. Therefore, all you can see of the right front wing is the edge. Hind wings are folded under front wings.

nae that are as long as their body and sometimes longer. One kind of katydid has antennae nine inches long! Shorthorned grasshoppers have shorter, rather thick antennae.

Katydids share certain other family characteristics. They all have long back legs that allow them to jump many times their own length. Their two front legs each have an ear. A funny place for ears? It works for the katydid, which uses its ears to find a mate. When a male katydid sings, females hear the song and join the male.

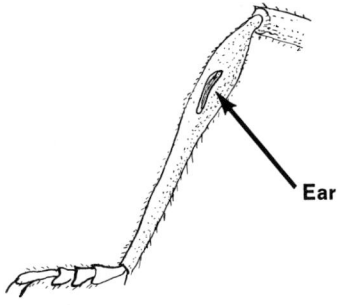

Front leg showing location of ear.

Katydids also have two pairs of wings. A stiff front pair is used for singing, and a collapsible back pair for flying. The back wings are kept folded under the front wings when the katydid is at rest. In many katydids, the back wings are so delicate and transparent, they look like beautiful fans when they are spread out for flying.

Wings of round-headed katydid spread out for flight and at rest.

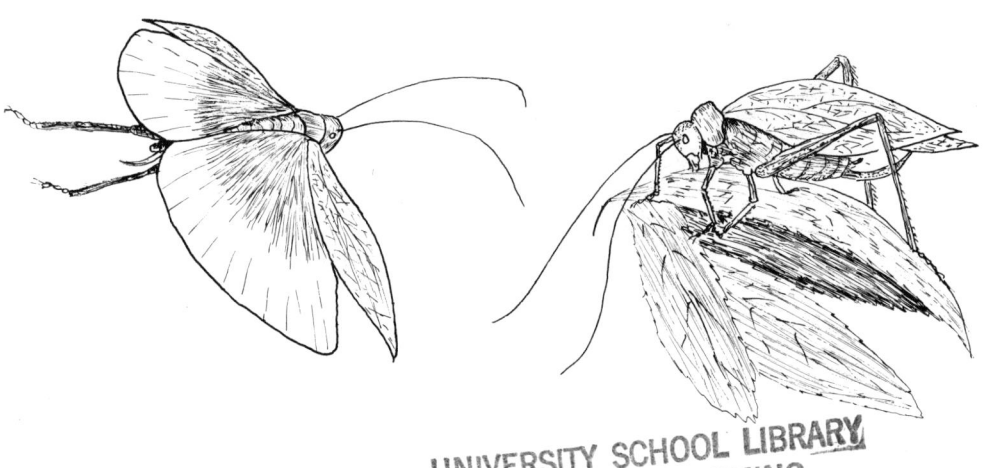

The shapes of front wings of six kinds of katydids.

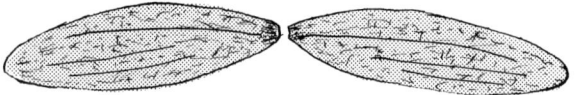

Oblong-winged katydid

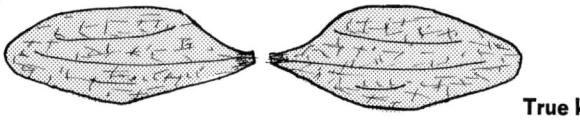

True katydid

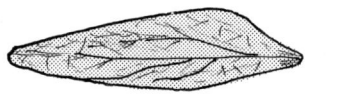 

Angular-winged katydid

Bush katydid

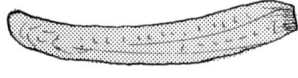 

Cone-headed katydid

 Mormon cricket

Some katydids have very short wings. They jump rather than fly.

The typical katydid color is green, but some are brown or black. A few tropical katydids are brightly colored, like the tropical plants on which they live. These colors make them as hard to see as green katydids are in the United States.

Female katydids have another characteristic: the *ovipositor*. This is a long, swordlike structure that sticks out of the rear of the body. The female uses it to lay its eggs in the ground or on leaves. Ovipositors vary in length and shape. One South American katydid has an ovipositor about three inches long so she can lay her eggs beneath the deep leaf cover on the forest floor.

Under the Linnean system, an animal family is divided into smaller and smaller groups. Two of these groups are called *subfamilies* and *species*. As the groups in the Linnean system get smaller, the animals in them have more and more characteristics in common. The species is the smallest unit in the system. Members of a species look almost exactly alike. They can mate and have young which can also mate and have young.

All this is true of the katydid family, too. In addition, each species of katydid sings a different song that attracts only a mate of its species.

There are thousands of different katydid species in

Some Katydid Subfamilies and Species

SUBFAMILY NAME	CHARACTERISTICS	WHERE FOUND
Round-headed or false katydids species include angular-winged katydid, bush katydid, Uhler's katydid	smooth, round forehead, large wings, green color	all over U.S. and southern Canada
Cone-headed katydids	cone-shaped head, large wings, green color	most of U.S. except western states
Meadow katydids	slender wings, green or brown color	in wet areas all over U.S. and southern Canada
Shield-backed katydids species include bog katydid, Mormon cricket	short wings or no wings, often black or brown color	western North America
True katydid*	long, stiff antennae, cupped wings, green color	east of Rocky Mountains

*The true katydid is just one species of the subfamily, but most of the others live in the tropics.

the world, about 255 of which live in the United States and Canada. Katydid species range in size from one-half inch to over five inches long. The biggest ones, which live in hot, wet areas of the world, look more like small birds than insects. The largest katydid in the United States lives in Florida and is about three inches long.

Occasionally, a strange-looking katydid is found that looks like a new species but is really just an old species with a new color. In 1967, a batch of bright pink katydids was discovered in a park in Westchester County, which is north of New York City. They were brought to The American Museum of Natural History in New York City. Alice Gray, a research assistant in the Department of Entomology, the department that deals with insects, identified them as the offspring of a common green katydid. The Museum put the live pink insects on display, and many newspapers ran articles about them.

A few weeks later, a man who had read an article about the pink katydids found a batch of yellow ones in a plant nursery in Westchester. They were also brought to the Museum and put on display.

These strange katydid colors are due to mutations. A *mutation* is a sudden and permanent change that produces an individual unlike its parent. Scientists think mutations occur fairly often in katydids, but few pink or yellow katydids are found because they are quickly eaten. They have lost their protective coloring.

David Nickle, a graduate student at the University of Florida, found a pink female katydid in Florida in 1972. He mated it with an orange male also found in Florida. The pair produced pink, orange and green offspring. By mating various members of these offspring, Nickle has been able to produce hundreds of pink, orange and yellow katydids, as well as the standard green ones.

David Nickle at work in the laboratory.

2

A Katydid's Life

Do you know Aesop's fable about the grasshopper and the ant? In the fable, the ant criticizes the grasshopper for singing all summer instead of saving food for the winter, like the ant. The grasshopper starves when the weather turns cold, but the ant lives on the food it has saved. The fable teaches a lesson, but it gives a false picture of the grasshopper.

Grasshoppers, including katydids, do not live very long, but their short lives have nothing to do with saving food.

Dr. Davidson Grove, an American entomologist (a scientist who studies insects) raised many angular-winged katydids, a common katydid in the United States. None of them lived for more than five months. This is a fairly typical life span for all katydids. In areas where there are hot summers and cold winters, they die as the weather turns cold. But even in areas where it is warm the year around, katydids do not live any longer than five months.

Like many animals in nature, katydids spend much of their time eating or looking for food. Young katydids, called *nymphs,* eat almost without stopping.

Nymph of tropical American katydid.

Most katydids are plant eaters, and their favorite part of the plant is the leaf. The kind of leaf they eat depends on what is available, but they have favorites. Dr. Grove's angular-winged katydids liked mulberry leaves best.

Dr. Stanley K. Gangwere of Wayne State University, Detroit, Michigan, watched captive katydids eating. Some species, he reported, chew a hole in the center of a leaf, and enlarge the hole until they come to a vein too tough to chew. Then they switch to another part of the leaf. Other species punch a hole in the skin of a grain such as wheat or barley and suck out the juice. When the grain is empty, they eat the skin, too.

Not all katydids are plant eaters. Some eat both plants and meat, and a few eat mainly meat. The meat eaters usually eat small dead animals they find, but some species chase their dinner. Many years ago, Henri Fabré, a French scientist, kept a female katydid in a cage. At first he fed her leaves, which she nibbled without much interest. Then he put a small shorthorned grasshopper in the cage.

Suddenly the katydid became excited. She chased the grasshopper, caught it with her front legs, bit it through the head, and ate it.

Fabré was shocked by this behavior, which was unknown to scientists at that time. But he was even more shocked later. He put a male katydid into the cage with the female, and the two mated. A little while later, the male died. The female, passing by his body, stopped, pulled off a leg and ate it. Today scientists know that cannibalism like this is fairly common among some katydids.

Being a creature that lives by itself, the katydid has to look for a mate. It uses a very good method: song. The katydid songs we hear are a sort of advertisement sung by the male to attract the female.

"What a male katydid says in his song is: "I'm a big strong male ready to mate," says Dr. Thomas J. Walker of the University of Florida, who studies katydids.

In some species, the female answers the male song with a little song of her own—usually a soft tick. The sound says: "Here I am." When Dr. Grove's angular-winged males heard the female of their species tick, they immediately jerked their body in the direction of the sound. They also lifted their front legs—the ones with the ears—as if to hear the female better. Then they flew toward the female, stopping, singing, and listening until they found her.

Not all katydids use sounds in this way. Dr. John D. Spooner of Augusta College, Augusta, Georgia, studied Texas bush katydids, and found that they use the male song and female ticks in three different ways. Sometimes the male sings, the female ticks, and the male comes all the way to her. Sometimes she moves part way toward him, and he comes the rest of the way. And sometimes he comes part way toward her, and she goes the rest of the way.

In many other katydids, the female makes no sound at all, but runs all the way to the male.

Once a male and female are together, the male pushes a bit of jellylike material out of his body and places it in the female's body. It contains sperm, the male reproductive cells. Inside the female's body, the sperm joins with the female's eggs. An egg that has joined with male sperm is called a *"fertilized egg."* Only a fertilized egg can produce a new individual. The eggs of a katydid look like tiny, flat seeds.

A day or two later, the female lays the fertilized eggs with her ovipositor. Some species lay their eggs in the ground, others on a twig or between the tissues of a leaf. The eggs laid in leaves eventually fall

The female ovipositing—laying the fertilized eggs with her ovipositor.

to the ground. In temperate climates, egg laying takes place in late summer or early fall.

When temperature and moisture are right, the little eggs swell and split, and out come the young katydids. In temperate climates, this happens in May or June.

Since most young katydids in an area come out of their eggs at the same time, the ground around an egg-laying site is full of nymphs at first. But the nymphs quickly scatter, because katydids keep away from each other except when they are mating.

The nymph looks almost exactly like its parents, except that it has no wings. To reach adult size, the nymph will undergo five or six *"molts"* in which it sheds its outer skin. A newly molted katydid is very

Completing the molting process.

soft, but its skin hardens within a few days. After each molt, the katydid is much bigger than it was before.

Most insects undergo this kind of growth, which is known as *"metamorphosis."*

Because they do not have wings as yet, katydid nymphs cannot fly or sing. But their long back legs enable them to leap. Dr. Grove measured the leaps made by some of his captive angular-winged nymphs, and found they could leap as far as 32 times the length of their body. When the nymphs mature, they do not jump nearly so far, because they depend on their wings to go long distances.

A winged katydid acquires its wings only after its last molt, which takes place in late summer in temperate climates. At first the new wings are very crumpled and limp but, after a few days they stiffen and smooth out. Katydids are not particularly good flyers. They fly only short distances, and most glide on the wind rather than flapping their wings up and and down like a bird. The true katydid can only glide downward from high branches to lower ones. To get back to the top again, it has to climb. The angular-winged katydid is one of the better katydid flyers. You can often see it flying up and down the streets at night, attracted by street lights.

When cold weather comes, the adult katydid becomes less active. It eats less, and the male song becomes slower and slower. Before winter, it is dead. But the eggs the female laid in the ground or on leaves will be safe until next spring, when the short life of the katydid will begin all over again.

3

The Song of the Katydid

Every time we talk about the sounds made by the male katydid to attract a mate, we call it a "song." But does this insect really sing a "song"? Yes. One definition of *song* accepted by scientists is "sounds organized in a certain way and always repeated in that way." By this definition, the sounds made by the male katydid to attract a mate *are* songs.

There are many differences, of course, between our songs and katydid songs. One big difference is that we produce melodies, or tunes, when we sing or play songs. Katydids do not produce tunes. They depend on changes in loudness, speed, and timing to produce variety in their music. One scientist compares katydid music to our drum music, because drums also use these changes to produce variety.

Another big difference between our songs and katydid songs is that we use both our voices and instruments to produce songs. Katydids do not have a voice; all their music is instrumental music. Their instrument? Their wings. When a male katydid gets ready to sing, he raises his front wings and moves them rapidly back and forth. By doing this, he rubs

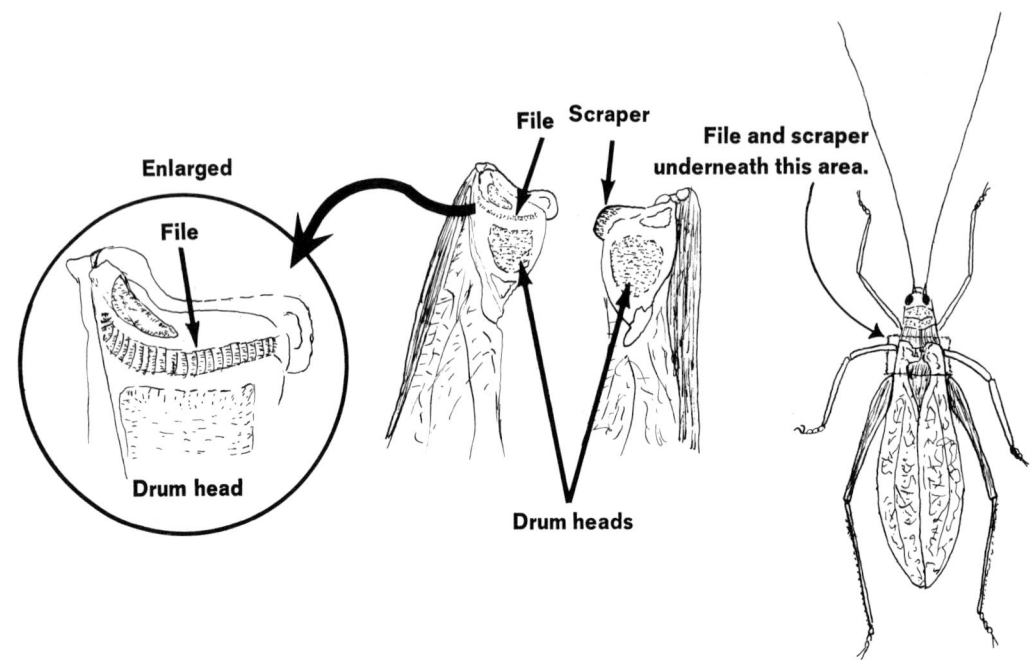

a *scraper* on one wing over a *file* on the other. The file has many tiny little teeth.

Next to the file and scraper on the wings are areas called *"mirrors"* that look like drum heads. They make the sound produced by the file and scraper louder.

What does the katydid song sound like? The song of each species is different, but people who have heard many katydid songs usually describe them as a buzz or rattle, often accompanied by ticks. Some scientists compare katydid songs to the sound of a

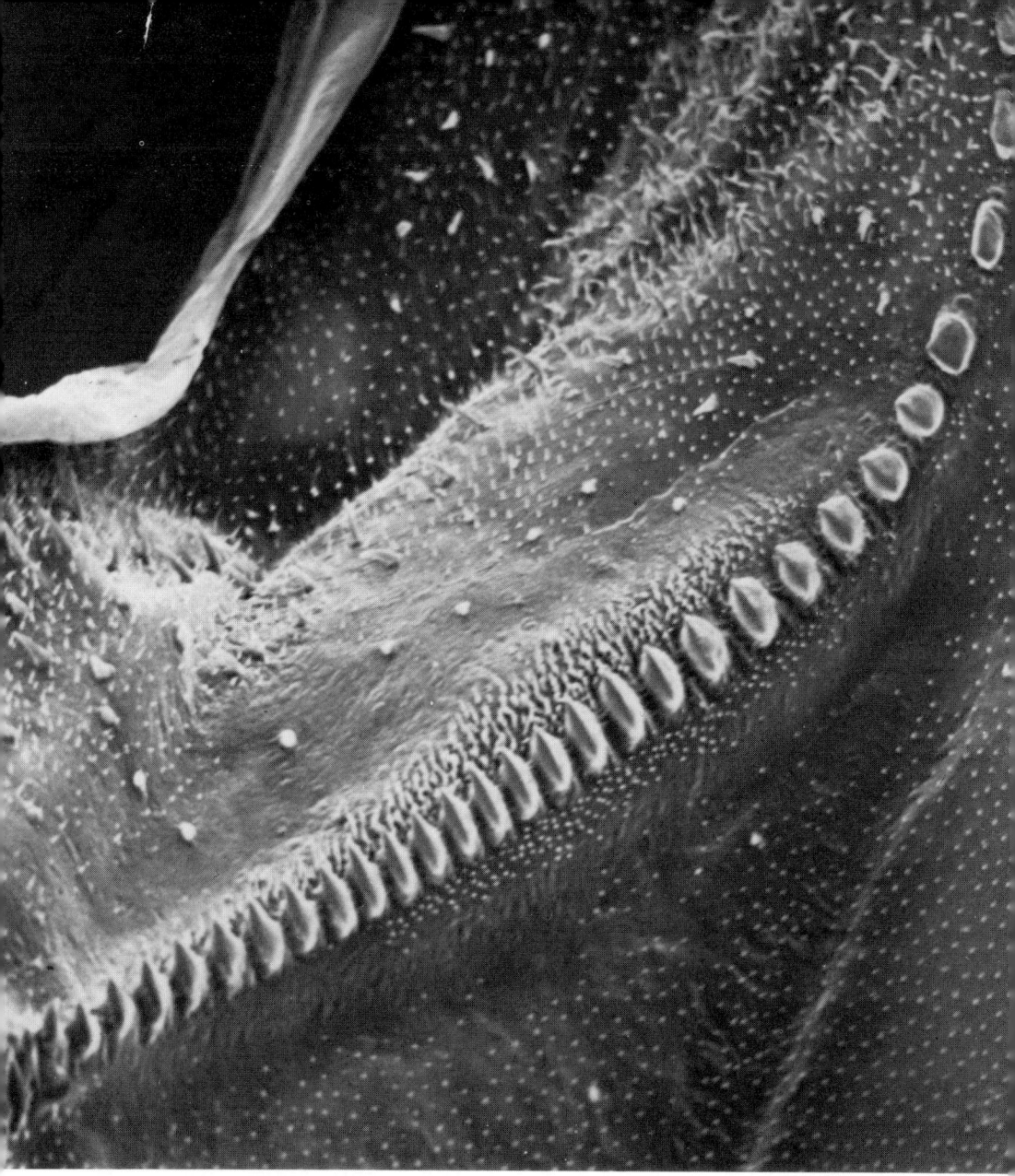

Enlarged view of teeth on file.

finger being drawn over the edge of a comb. Dr. Ross Hutchins, an entomologist who has written many books on insects, says most katydid songs sound like "a pocket watch being wound."

The most famous katydid song gives the whole family its name: "katy did." Some scientists feel, though, that this song really sounds much more like "ka ki-KAK" than "katy did." In any case, the true katydid sings "ka-KAK" in some areas, while in others it sings a longer song.

Some katydid songs are rather soft, others are loud, but almost all can be heard clearly in a quiet area. Indoors, some katydid songs sound amazingly loud. Dr. Glenn K. Morris, an entomologist at the University of Toronto, Toronto, Canada, captured some katydids one summer, and kept them in a house where he and his family were staying. Their song was so loud, it kept the family awake. Dr. Morris measured the song of this species at 100 decibels in the field, which is as loud as a car passing close to you on the highway.

Most katydid songs are not this loud, and many sound quite pleasant to our ears. In the eighteenth and nineteenth centuries, people in Germany used to keep male katydids in their houses so their song could be enjoyed. The katydids lived in colorful little cardboard houses called "hopper houses."

A Canadian entomologist, Dr. D. K. McE. Kevan, who has researched these hopper houses, says you could buy them flat and fold them together or buy an assembled house with a katydid already inside. An eighteenth century print shows a katydid seller carrying a number of hopper houses. The sellers used to cry: "Grasshopper in the cage, worth one shilling!" Dr. Evans points out that this grasshopper was always the katydid, as it was considered a far better singer than the shorthorned grasshopper.

Selling grasshoppers in hopper houses.

But we do not hear katydid songs the way katydids hear them. When a katydid makes a sound, it produces trembling movements called *vibrations* in its wings. These vibrations cause a sound wave to move

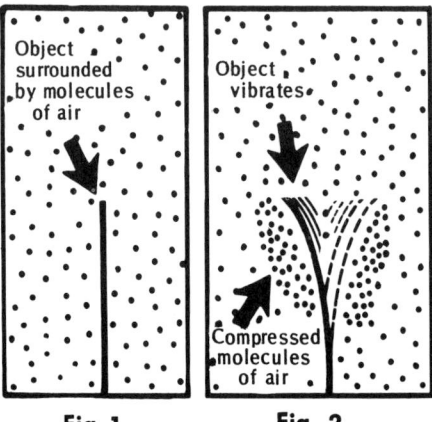

Fig. 1   Fig. 2

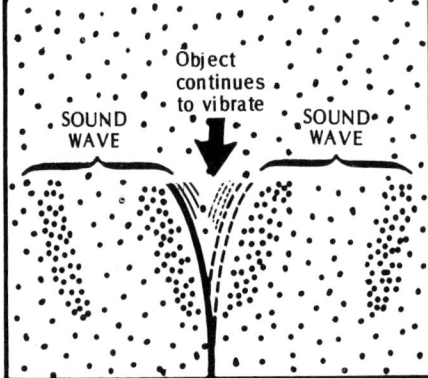

Fig. 3

Sounds are made when an object vibrates rapidly. Fig. 1 shows an object surrounded by molecules of air, represented by the dots. In Fig. 2 the object begins to vibrate, causing nearby molecules to bunch up, or be compressed. Since air has elasticity, the bunch of air molecules quickly expands again. But this compression and expansion causes other molecules farther away to be pushed together, as in Fig. 3, and this second group of molecules, now in motion, causes a third disturbance even farther out, and so on. This sequence of compressing-expanding groups of molecules forms a "wave" of sound. The frequency of this wave is the number of times per second that the object vibrates to form the compressed group of molecules.

through the air, vibrating at the same speed as the wings. If the vibrations are fast, we hear a high sound, if slow, we hear a low sound.

The number of times per second an object vibrates is known as its *frequency*. Many katydids make sounds with very high frequencies. Some are so high we cannot hear them, although katydids can. We call sounds like this *"ultrasonic."* The only way we can hear these sounds is to slow them down on special sound equipment.

One piece of equipment that has played a very important role in studying katydid songs is the *oscilloscope*. It enables scientists to look at sounds on

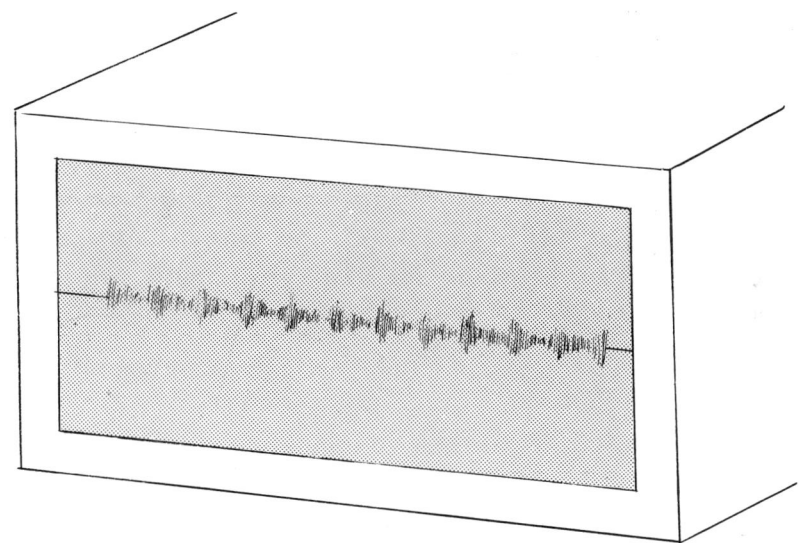

Oscilloscope recording of a katydid song.

a screen. The sounds appear as a pattern of lines that can be read. By using the oscilloscope, Dr. Thomas Walker and Donald Dew of the University of Florida found that the song of one katydid, Uhler's katydid, has four different sounds. This is more sounds than the song of any other insect.

The song of Uhler's katydid goes like this:

    buzzzzzzzzzZZZZZZZZ
    RATTLE–RATTLE–RATTLE
    CHHH–CHHH–CHHH–CHHH
    tick–tick–tick–tick

Each kind of sound is made by a different movement of the wings. And the insect sings the four sounds in the same order each time. The whole song takes eight seconds or a little longer.

Does Uhler's katydid have a more complex scraper and file than other katydids? Surprisingly, the answer is no. Uhler's katydid has simpler singing equipment than some of its relatives. But the way its scraper and file are made does not seem to influence the kind of song it sings.

Outside his laboratory, Dr. Walker has often observed another kind of katydid singing: chorusing. A *chorus* is a group that sings together. Although each katydid usually keeps to itself, often a large number

Some Well-Known Katydid Songs

SPECIES	SOUND	RANGE	WHEN AND WHERE SONGS ARE SUNG	CHARACTERISTICS
angular-winged katydid	chirp (song I) tick (song II)	eastern, southern and western U.S.	high in trees at night	chirp repeated at well-spaced intervals, tick in rapid series
oblong-winged katydid	shrie–e–e–e–k	eastern U.S., southern Canada	bushes and vines at night	very low rasping sound repeated several times
fork-tailed bush katydid	zeep	all of North America	low shrubs at night	soft, high-pitched sound repeated several times
true katydid	ka ki–KAK or ka ki–ka–KAK or ka–KAK	eastern U.S.	trees at night	sound repeated about 60 times per minute
swordbearing conehead	tick or zip	east of Rocky Mountains	low shrubs and grass at night	sound repeated indefinitely
common meadow grasshopper	long buzz followed by ticks	east of Rocky Mountains	moss and weeds during day and night	low sound lasting two to three seconds
Orchelimum gladiator (another meadow grasshopper)	buzz followed by ticks	northern U.S., southern Canada	grasses and low shrubs during day	ticks cannot be heard unless insect is very close

NOTE: These songs sound different to each hearer, so you may not hear songs exactly like those described.

of katydids of one species will station themselves a little apart from each other in an area and sing together. They keep such perfect time that it sounds like one insect singing.

If a new insect arrives and begins to sing early, the whole group pauses for a moment. Then the chorus starts up again with the newcomer singing along with the others.

In some species, says Walker, hundreds of individual chorusing katydids will divide themselves into two choruses. First one chorus sings, then the other answers it. In these species, katydids will also answer each other when just a few katydids are present.

One factor that affects all insect singing is temperature. A katydid that sings fast in warm weather sings more and more slowly as the weather becomes colder. On chilly October nights towards the end of the katydid's life cycle, they sound like a record that has been slowed down.

◀ In the field, Dr. Walker uses a portable dish antenna for recording katydid songs. With this equipment, he can pick up sounds from a long distance.

4
A Katydid Laboratory

It was almost midnight one summer evening, but Dr. Glenn Morris was still in his laboratory. He sat in a small room that had been soundproofed so that no outside noise could be heard. On a table in front

Dr. Morris and his assistant with the arena. Dr. Morris is playing the male katydid song through a loud speaker to attract the female.

of him was a circular structure that looked like a miniature sports arena.

A female bog katydid, a small brown insect with green stripes on her sides, sat on a post in the arena. The bog katydid gets its name because it is often found in low, wet areas called *bogs*. It is one of the few katydids that lives in the far north of Canada. Dr. Morris collected this female and other bog katydids in western Canada.

He switched on a speaker mounted on the arena wall. The speaker was connected to a tape recorder in a room next door. Buzzzzzzzzzzz. The sound that filled the room was the song of the male bog katydid. Dr. Morris knew that the males of other species of katydids use their song to attract females. But was the same true of the bog katydid? And if it was, how would the female act?

As soon as the buzzing sound was heard, the female became very excited. Antennae moving rapidly back and forth, she climbed down from the post and zigzagged across the arena floor. When she reached the wall, she climbed up it and stopped in front of the speaker. Meanwhile, Dr. Morris was drawing the female's path on a piece of paper. When the bog katydid stopped, he turned off the speaker and put her back in her cage.

For the next six nights, Dr. Morris put one female

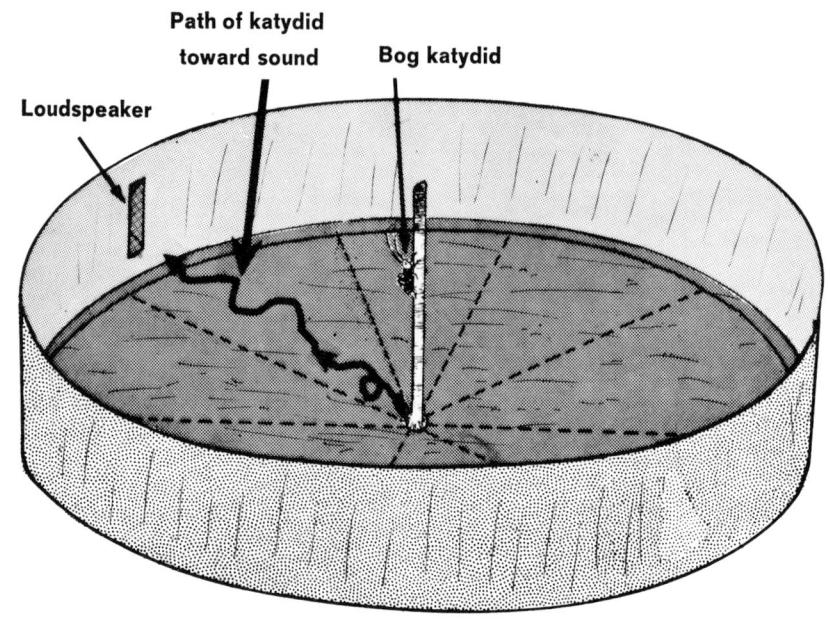

Path taken by a bog katydid in the Morris arena.

bog katydid after another through the same experiment. He used 11 females in 204 trials.

In almost every case, the female that heard normal male song coming from the speaker dashed over to it. Just as Dr. Morris had expected, male bog katydids use song to attract females. The experiment also showed that the female bog katydid needs only the sound to find the male. She does not need to smell him or see him.

However, the most suprising result of the experiment was that females pay more attention to some sounds than others.

Male bog katydids, like some other katydids, sing a song with two kinds of sounds: audio and ultrasonic. *Audio* sounds can be heard by human ears; ultrasonic sounds cannot. To hear ultrasonic sounds, we have to slow them down on special sound equipment.

When Dr. Morris played a song from which he had removed all audio sounds, the females ran to the speaker half the time. When he played a song from which he had removed all ultrasonic sounds, the females paid no attention to it at all. This means that the ultrasonic song part allows the female to find the male, but the audio part does not.

Dr. Morris thinks the ultrasonic part is a sort of signal or beacon to guide the female to the male, while the audio part simply makes her excited.

What kind of signal does the ultrasonic part give the female? Dr. Morris does not know, but he has a theory that might answer this question. In science, a *theory* is a possible explanation based on evidence. Dr. Morris' theory is founded on the fact that sounds lose their energy as they travel away from the source, and become fainter and fainter until finally they cannot be heard at all. Ultrasonic sounds lose their energy much faster than audio sounds. This means that as the distance from a bog katydid singer increases, the ultrasonic sound gets weaker faster than the audio. These differences could help a female katydid judge how far away she is from a male, Dr. Morris believes.

Male bog katydids also have another song, the "courtship song," which they sing when they see the female. This song has no ultrasonic sounds. Dr. Morris thinks that this supports his idea that the audio part of the song excites the female instead of guiding her.

Glenn Morris is one of the few scientists in the world who studies the behavior of katydids, particularly their songs. "I'm interested in two things," he says. "Collecting all the katydid songs I can find, and taking songs apart to see what the various parts contribute to them." When he is not teaching classes, he looks for katydids and records their songs.

He brings some of the katydids back to his laboratory to use in experiments. The katydids he finds in cold climates are kept in a small, chilly room that recreates the temperature they experience in the wild. The insects get 12 hours of light and 12 hours of darkness. "Hear that thing that sounds like a train?" he asks as he shows a visitor the room. "That's the bog katydid. I brought them back from Thunder Bay, at the north end of Lake Superior."

When he wants to record a song or run an experiment with a katydid, he takes the insect from the cold room to the soundproofed room. The most im-

Dr. Morris records in the field by using a microphone which can only pick up sounds close by.

portant and expensive items in the room are the tape through another part of the oscilloscope."

"The oscilloscope shows changes in the loudness of a sound in a period of time," explains Morris. "This is the song of the katydid I'm studying. If I didn't know what species this is, I could tell by looking at the pattern on the screen. When we want to make a permanent record of a song, we run a strip of paper through another part of the oscilloscope.

The oscilloscope has a small green screen that looks like a miniature television screen. Below the screen is a control panel with a number of dials. When Dr. Morris plays the tape of a katydid song, a moving line appears on the screen. The line becomes a pattern that remains on the screen when the song is over.

"One thing we can do with the tape recorder is slow down a song," he continues. "Ultrasonic songs are sung too fast for us to hear, but by playing them back slowly out of the recorder into the oscilloscope, we can see what they are like."

Not all Dr. Morris's experiments take place in the laboratory. Some of the katydids he is interested in are meadow katydids. Most of the members in this subfamily of katydids that are active during the day do not fly. Two species live in the Toronto areas, and Dr. Morris often watches them in nearby fields. He has found that singing males keep a certain distance between themselves. If one gets too close to another, they wrestle until one runs away. Dr. Morris has seen these "wrestling matches" so often that by now he can set one up for his classes and visitors.

Dr. Morris sometimes uses a net to catch katydids.

First he collects a number of males from one area, and puts them in another area where males of the same species are singing. The males sit on grasses about a foot above the ground. Before long, one of the "foreign" males begins to sing. Almost immediately, a "resident" male begins to hop toward the foreigner. The resident passes a nonsinging foreigner but pays no attention to him.

When the resident reaches the singing foreigner, he rushes up the grass blade to which the foreigner is clinging. The two meet, put their chests together on opposite sides of the grass blade, and wrestle. Kicking and biting are part of the wrestling match. One male falls down to a lower grass blade, but he climbs right back up again and continues to wrestle. Then both males tumble to a lower blade, where they keep wrestling. Finally, one male hops away.

The other promptly climbs up high on a grass blade and starts to sing again. The males seldom hurt each other in these wrestling matches, according to Dr. Morris.

"Most people think male katydids sing only to attract females," he says. "But for these two species of meadow katydid, the male song also makes other males keep the right distance from each other."

5

Katydid Pests

It was a sunny afternoon in 1848, near what is now Salt Lake City, Utah. The area had been settled the year before by Mormons, members of a religious group. A Mormon farmer working in his wheat field looked up and saw what seemed to be a long, wide black stream flowing slowly toward him.

Dashing up to it, the farmer saw the stream was made up of the familiar black insects which Indians in the area ate as food. The Mormons called them crickets, although they were really a flightless katydid.

There were thousands of the crickets in the stream —no, millions! It was so long the farmer couldn't see the end of it. He stamped on some of the insects, but the stream flowed around his feet and moved toward his wheat field. When the insects reached it, they began to eat the young wheat. Soon there were so many insects in the field that it was completely covered with them. By dark, the field was eaten to the ground.

The next day, the insects moved on to another field —and then to another. The little community was in a panic. This was their first crop. If it was eaten by in-

Crickets jumped into a moving stream and collected on a log that was placed in the water to prevent them from moving downstream.

sects, the settlers might starve, since they were a thousand miles from the nearest settlement. The Mormons began to fight the insects. Some struck them with brooms and clubs, while others dug ditches around their fields and filled them with water. Some insects fell in the ditches and drowned. One young girl drove sheep across her family's fields to trample the crickets.

But there were too many crickets. It looked as

though the Mormons' crops were doomed. Then, a huge flock of seagulls suddenly appeared overhead. They came from islands in the Great Salt Lake. The birds swooped down and began to eat the crickets.

"In two days, the black plague was destroyed," wrote one Mormon.

Years later, the residents of Salt Lake City erected a

Statue erected to the gulls which destroyed the Mormon crickets.

statue to the gulls. You can still see it in Temple Square. The wording on the statue reads:

But that was not the last of the Mormon crickets, as the black insects came to be called. A member of the shield-back subfamily of katydids, the Mormon cricket is still an agricultural pest in the western United States. Since 1848, there have been a number of serious *outbreaks* of Mormon crickets. An outbreak is a

A male shield-back, the Mormon cricket.

sudden increase in the number of insects, which then begin to move as a group. The most recent outbreaks occurred in the 1930s.

Today, Mormon crickets are less of a threat to American farmers. *Pesticides*—chemicals used to kill insects and other pests—are spread by plane in mountainous areas where Mormon crickets breed. The pesticides are usually in a food, such as wheat bran. Now, however, the federal government has banned these pesticides because they have harmful effects on livestock and wild animals. New pesticides have been developed that are not harmful, but some agricultural experts are afraid they will not work as well as the old ones.

Another way to catch crickets was to use a metal trap. As they moved into the trap, the crickets piled on top of each other, suffocating to death.

"I think we'll see a lot more Mormon crickets in the future," says Dr. John Henry, an entomologist at the Rangeland Insect Laboratory in Bozeman, Montana.

Already, the population of Mormon crickets is building up again in some areas. Idaho has a particularly large number of them.

Why does the Mormon cricket band together to become an agricultural pest when most of its katydid relatives are solitary?

Scientists are not sure of the answer, but they do know that many more Mormon crickets are born in some areas during certain years. When temperature and wind speed conditions are right, the crickets all begin to move in the same direction, or *migrate*. Since they do not fly (their wings are very short), they move slowly. Even so, a band of Mormon crickets can cover up to a mile a day, eating as they go.

Mormon crickets eat both crops and wild plants that grow where sheep and cattle graze. Their favorite parts are the flowers and seeds, but they often eat the plant right down to the ground. One observer saw Mormon crickets eat a stand of five-foot-high mullein plants with stems two inches in diameter. When the insects had eaten the mullein to the ground, they began eating as much of the stems as they could reach below ground.

The Mormon cricket has a relative, the coulee cricket, which is also an agricultural pest in the western United States. Like the Mormon cricket, it is a flightless, shield-backed katydid which migrates in large groups. The coulee cricket has a smaller range, and so it has never been as much of a problem as its relative. Today the coulee cricket is kept under fairly good control by pesticides.

A few other katydids are agricultural pests in small areas. One of them is a bush katydid that lives in California. Bush katydids are members of the round-headed subfamily. In some years, when a large number of bush katydids are born, they begin to eat plants such as young oranges and roses. They do not devour the plants entirely, but they do damage them.

For the most part, though, katydids do not do much harm to human beings. In the early West, Mormon crickets were a large part of the diet of some Indian tribes. The Indians roasted the insects, pounded them into a meal, added water, and baked the mixture. Mormon cricket soup was another favorite Indian dish. Or sometimes the Indians ate the soft body of the insects raw.

Insects are rich in protein, so eating Mormon crickets helped keep the Indians healthy. The practice may also have killed enough Mormon crickets to reduce damage to plants before the settlers pushed the Indians out.

6

A Katydid Collector

Dr. David C. Rentz of the California Academy of Sciences in San Francisco, California, is a biologist and katydid collector. In the 20 years he has been collecting katydids, he has found at least two dozen new species. In the United States, he has collected katydids in areas ranging from the deserts of northern Nevada to an area near Malibu Beach not far from Los Angeles.

Dr. Rentz on a collecting trip on Pt. Conception, California.

Dr. Rentz is particularly interested in the shield-backed subfamily because they have interesting habits and are not well-known. The reason why they are little known, he explains, is that shield-backs are even more secretive than other katydids. Also, many live in small areas. Shield-backs have tiny wings that are almost completely covered by a shield-like horny plate on their backs. Most shield-backs are blackish or brownish. Some of the shield-backs Dr. Rentz has collected are not new species, but they are rare, nevertheless.

"This particular shield-back comes from near Reno, Nevada," he says, picking up a box with a big, grayish insect inside. It is pinned to a styrofoam block. "See how short the wings are? Shield-backs can't fly with these little wings, but most of them can sing very well. You can hear this one 50 feet away.

"There were some old specimens of this species in collections, but I couldn't find any live ones when

(a)

Some of the katydids collected by Dr. Rentz.
(a) Opposite page. This is a female shield-back. Note tiny wings and long ovipositor used in egg laying.
(b) Adult male shield-back from Africa. Wings are concealed under the shield.
(c) A tropical American katydid from the rainforest in Costa Rica.

I first visited Nevada in 1968. But in the winter of 1969, there were heavy rains there, and when I went out the following summer, these katydids were just all over the place. What happens, apparently, is that the eggs of some of these insects remain in the ground until a year comes along with just the right amount of moisture. Then they come out."

Dr. Rentz has also collected katydids in South America, Central America, and South Africa. On his latest trip, to South Africa, he took along his usual collecting equipment. It includes a butterfly net, test tubes, "killing jars," a number of cigar boxes, and a supply of cellucotton. The *killing jar* contains a poison that kills the insects when they are put in it. The nets and test tubes are used to capture insects, and the cigar boxes and cellucotton hold dead specimens. A *specimen* is an animal or thing typical of its group.

Dr. Rentz found many of his specimens by driving at night down little-traveled roads in the South African desert. He uses the method in deserts here, too. "Most of these katydids go out at night looking for food, and a lot of them are ground dwellers, so they move across the ground," he says. "We have difficulty seeing them on the desert itself, but you can see them easily on the road."

As he collects, Dr. Rentz makes notes on where he finds specimens and on their behavior. He uses this information later in articles he writes for scientific journals.

On his South African trip, Dr. Rentz found that he had arrived at the wrong time to collect some of the adult katydids he wanted. The katydids were still nymphs. So he captured the nymphs, put them in jars with tops made of screening, and brought them back to the United States. He kept the nymphs under light in his laboratory, and fed them Purina dog chow, a favorite food of many captive katydids. Before long, most of the nymphs became adults.

He also brought back many dead katydids in the cigar boxes. Each one had a label next to it saying where it was collected, the date, and Dr. Rentz's name. As soon as he reached his laboratory, he put the dead insects in a *humidity chamber.* This is a box with a wet sponge and some moth crystals to keep mold from forming. After the katydids had been in the humidity chamber for eight hours or so, their bodies absorbed enough moisture so that their joints could be moved easily without breaking. Dr. Rentz removed them from the chamber, pinned them, labeled them, and put them in the collection.

The katydids went to the insect collection of the Philadelphia Academy of Natural Sciences, where Dr. Rentz worked at the time. The Academy has the largest collection of grasshoppers in the world—about two million specimens. Some are over 100 years old. The grasshoppers are kept in glass-topped wooden trays placed one above the other in tall metal filing cabinets. The big room in which they are kept smells of moth balls, used to fight a beetle that eats dead insects.

An assistant looking at the insect collection of the Philadelphia Academy of Natural Sciences.

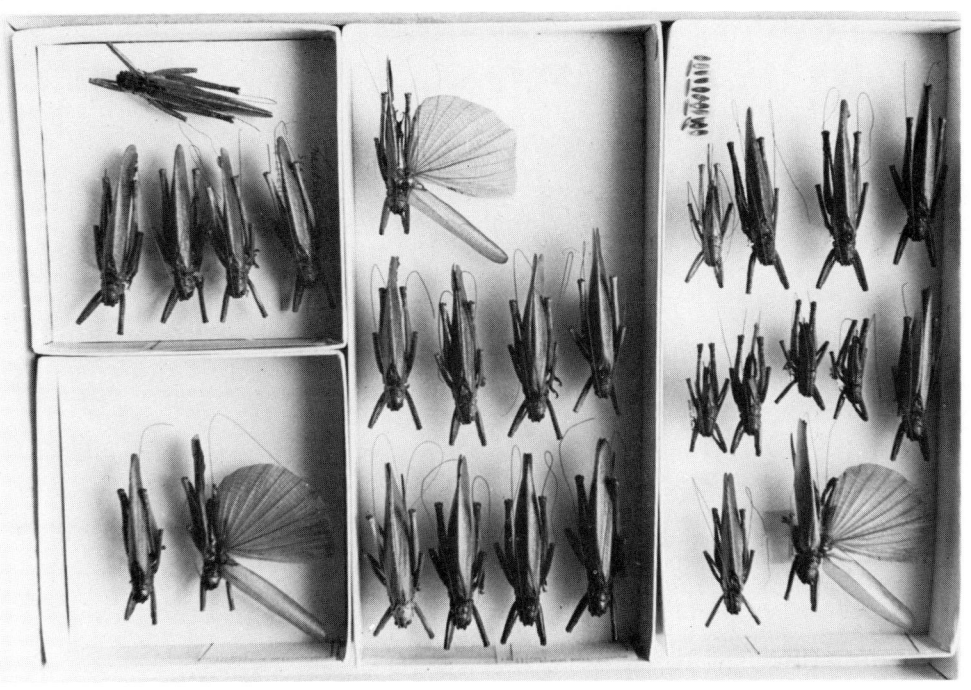

A drawer of katydids from the collection.

Before Dr. Rentz left the Academy, he described some of the more unusual katydids in the collection to a visitor.

"Look at this one", he commented, pulling out a drawer. The katydid he was talking about was a long, heavy-bodied insect the size of a sparrow. "Its body is about five inches long and its wingspread about four inches. It comes from Malaysia. But there are others that are almost as big." Dr. Rentz pulled out other drawers to show huge katydids from the islands of Borneo and Java in Indonesia. Although these big katydids are fierce-looking creatures, they are peaceful leaf eaters.

"This is one of our rarest katydids," continued Dr. Rentz, sliding out a drawer in another section of the room. Inside was a grayish katydid with spines on its back and gauzy, spotted wings. "This is a lichen mimic. The body and wings look like the lichen on which it lives." He opened another drawer. "And this one is a leaf mimic—the best leaf mimic in the world. Notice how the wings have what look like little torn places, as if they had been chewed on by insects. But each animal has the holes in exactly the same places. They are not holes at all." Both these mimics come from South America.

"This is a research collection," Dr. Rentz went on. "It's almost like a library. People write in for material, and the Academy sends it to them and they keep it for years, sometimes. The Academy sends out thousands of insects a year." Researchers need these insects to study while they are describing a new species within a subfamily.

This research, which is known as *classification,* is an important part of Dr. Rentz's work, too. For instance, at the present time he is describing some of the many shield-backed katydids and arranging them within their subfamily. When he is finished, he will publish a paper on the subject in a scientific journal.

Classification is a job for a specialist, but collecting katydids can be done by amateurs. Sometimes amateurs even find new species. In 1974, a high school boy named David C. Lightfoot who lives in Corvallis, Oregon, sent Dr. Rentz a new species of katydid. He had found it near Abert Lake in Oregon. Now he and Dr. Rentz may write a scientific paper together describing their katydid.

"There's a very good chance that an amateur can find a new species if he lives in an out-of-the-way place in the West," says Dr. Rentz. "A lot of collecting has been done in the East but not too much in the West."

7

The Katydid's Relatives

The katydid has five orthopteran relatives. All of them have the typical straight wings of this group, but not all look alike. Crickets and shorthorned grasshoppers, including the locusts, look like the katydids. All of these insects sing. The non-singing orthopterans —the cockroach, walkingstick, and praying mantis—do not look like a katydid.

There are a few clues you can use to tell a katydid from a shorthorned grasshopper or cricket. One clue, which we have already mentioned, is antennae length. Shorthorned grasshoppers have short, thick antennae. Longhorned grasshoppers or katydids have long, thin antennae. Crickets have long, thin antennae too, but their body color is usually black. Most katydids and grasshoppers are green.

If you look at an orthopteran very closely, you can tell katydids and crickets from shorthorns by the position of their ears. In katydids and crickets, the ears are near the knees on the front legs. Shorthorns have them on the side of their bodies, near the upper end of the first pair of legs. The orthopteran ear looks like a small slit or hole.

Where do locusts fit into this picture? Years ago, a Russian researcher discovered that when shorthorned grasshoppers of some species are crowded together into a small area, they soon change into another kind of grasshopper—the locust. Locusts, like Mormon crickets, have a strong tendency to migrate. Since locusts fly, and fly well, they can do much more harm than their crawling cousins, the Mormon crickets.

There are about six hundred species of shorthorned grasshoppers in the United States and Canada. Only half a dozen damage crops, mostly in the western states. Since the 1930s, there have been no serious outbreaks of locusts in the United States. The worst locust outbreak took place in the 1870s, when the Rocky Mountain locust suddenly descended on western farms.

The Rocky Mountain locust has not been seen since 1892, but it still exists in a nonmigratory form. Probably so many of its breeding grounds have been destroyed that it cannot build up a large enough population to migrate.

In Africa and Asia, locusts are still a serious menace to crops, as they have been for thousands of years. The Bible's Book of Exodus describes a swarm of locusts that ate all the crops in Egypt. The locust that did this damage may well have been the desert locust, the most destructive locust in the world. It is a big insect about six inches across the wings, and a strong flyer.

The cricket is a harmless creature compared to the locust, but it does nibble on crops and paper, and sometimes on cloth. The best-known crickets in the United States are the house cricket, the field cricket, and the snowy tree cricket. The first two are both black and an inch or less in length. The house

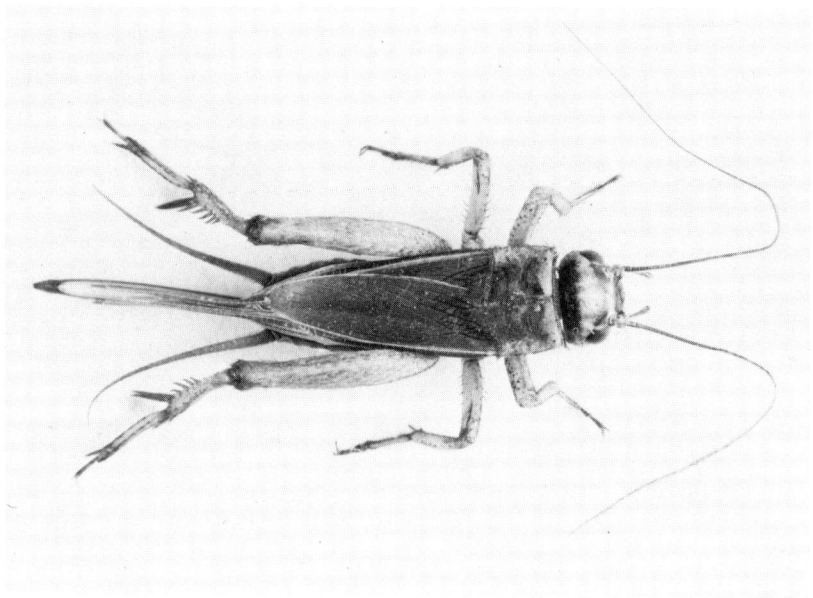

House cricket.

cricket, which likes warmth, is usually found indoors in the fall, but the field cricket, too, often comes indoors. One way to tell them apart is by their song. The house cricket chirps faster than the field cricket. The snowy tree cricket is a pale green insect about a half inch long. It usually sings from low shrubs in the evening.

Most insects are not great favorites of human beings, but the cricket is an exception. In the British Isles, a cricket singing near the fireplace used to be considered lucky. The Chinese kept singing crickets as pets in small cages. Today, in many oriental countries, crickets of the same size are matched in fights that continue until one insect is dead.

If the cricket is our most popular insect, the cockroach may be the least popular. Many people say it leaves a mess in their kitchens, and carries disease. Entomologists admit the cockroach secretes a fluid that gives food a bad odor, but they say the cockroach has never been shown to carry disease. John Kethley of the Field Museum of Natural History in Chicago claims the cockroach is a very clean animal.

"They spend hours grooming themselves and are extremely fastidious," he says.

Unlike its singing relatives, the cockroach has rather short back legs. To get away quickly, roaches depend on a quick dash rather than flying or jumping. Some roaches, however, do fly. The usual color of cockroaches is brown or black. There are about 55 species of cockroaches in the United States. The most common are the German, the Oriental and the American.

The number one cockroach as far as numbers go is the German. This is the half-inch long, shiny brown roach people often find in cupboards and drawers. The American cockroach, which is medium to dark brown grows up to two inches long. It lives in basements and the first floor of buildings. Sometimes you see it running along a city sidewalk. The Oriental cockroach is blackish and a little smaller than the American. It, too, lives on the lower floors.

The walkingstick is the world's biggest insect. Some tropical species reach 15 inches in length. The biggest one in the United States is seven inches. "Walkingstick" is a good name for these insects, because they are long and thick, like a stick. The color of the walkingstick changes with the seasons. In spring, it is green, in fall, brown or gray. When a walkingstick is sitting on a tree or bush, it is hard to find, making its stick disguise good protection.

Some walkingsticks are pests. According to a study by entomologist Charlie Rogers of the U.S. Department of Agriculture, walkingsticks ate many acres of

Walkingstick.

forest trees in 1972 in Arkansas and Oklahoma.

The only orthopteran that does *not* eat vegetation is the praying mantis, which is strictly a meat eater. Its "meat" is insects. The katydid is one of its favorite meals. The mantis is a big insect, the most common species in the United States measuring 2½ to 3 inches long. The body is hinged so that the front part can be carried upright, and the head can turn from side to side. This helps the mantis find and capture prey.

But its long front legs help it most. Like most insect legs, they have three sections; the last two have sharp barbs. At rest, the mantis holds these last two sections against each other, so that the barbs are inside the fold. The name of the praying mantis comes from this particular position, which does make the insect look as if it is praying.

In reality, though, the mantis is waiting for prey. When a live insect draws near, the body rears up and the barbed arms slowly unfold. Then the arms flash out, grab the insect, and fold up again. The prey is held securely by the barbs while the mantis eats it alive. The female mantis is considerably larger than the male, and a number of scientists have observed her dining on the male after he has mated with her.

Farmers like the mantis because it eats a number of harmful insects. Today, many ecology-minded gar-

Praying mantis.

deners are buying praying mantises to control pests rather than use harmful chemical pesticides. The mantis makes an interesting pet, too. It is hardy and survives many months in captivity on a diet of live, medium-sized insects like crickets.

8

How to Find, Catch, and Keep Katydids

The scientists we have been reading about study katydids as part of their work, but katydids are fascinating subjects for home and school study, too. In the field, you can do several experiments with katydids. Dr. Thomas Walker has worked out an experiment in which you can teach katydids to count!

Some true katydids, as we saw earlier, sing a song that sounds like "ka ki-KAK", but others sing shorter or longer songs. If two katydids are present, first one sings, then another answers it. You can take advantage of this behavior to make a true katydid count. If the true katydids in your area sing "ka ki-KAK", wait until one finishes, and then imitate the song with your mouth but make it longer: "ka ki-ka-KAK".

The next singer may sing a longer song like yours. If this happens, wait a while and then do still a longer song: "ka ki-KAK ki-KAK". Again, do your imitation just after a katydid has finished his song. The answering katydid may sing the same song. Wait a while longer, then try a short song: "ka-KAK." The next katydid to answer may cut down the length of his song.

Another experiment, also suggested by Dr. Walker, involves male and female katydids. In species in which females answer males, you may be able to get a male to fly to you by imitating a female. The best-known species in which females answer males are the angular-winged katydid and the many species of bush katydids. All are common in the United States. Check the chart on page 37 for a description of the songs of these species, or use a record of insect songs.

The record, *The Songs of Insects,* issued by Houghton Mifflin Company for the Cornell University Laboratory of Ornithology, includes the song of the angular-winged katydid and the Teaxs bush katydid. It can be ordered from the Laboratory of Ornithology, Cornell University, Ithaca, New York 14850. Price is $7.75 plus sixty cents for postage.

To imitate a female, simply make a little ticking sound. You can do so by clicking your finger nail, by making a sound with your mouth, by hitting a dime against a belt buckle, or by twanging a single tooth on a comb. Wait until the male or males stop singing, and then make your tick. In bush katydids, the time between the end of the male song and the female tick is within a second.

If the male thinks you are a female katydid, he

will fly a little way toward you and repeat his song. Each time he finishes singing, make your tick. It usually takes 15 or 20 minutes for a male to get close enough for you to see him. Besides the singing male, you may also attract a number of nonsingers looking for a female.

You can study katydids in your home, too. Before you go on a katydid-hunting expedition, get several widemouth jars or pint-size ice cream containers and punch holes in the tops. For catching katydids in tight places, such as between branches in dense shrubbery, a widemouth test tube is suggested by Dr. Rentz. If you can't get such a test tube, the casing an aquarium heater is mounted in will do just fine. It can be purchased at many stores that sell tropical fish.

The most likely place to find day-singing katydids is a weedy field with low vegetation. Holding a jar or container in one hand, select a single specimen to approach. Dr. Morris suggests you take a friend along and use a procedure called "triangulation" to locate the insect. Stand apart from your friend and with an arm held out straight, point to where you think the song is coming from. Your friend does the same. Then both of you move slowly along the path you have indicated.

Hopefully, your paths will cross at the point where

Equipment for a katydid-hunting expedition.

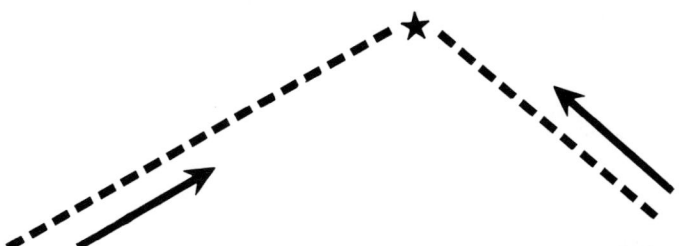

The triangulation method needs at least two persons to join in the search. Each person walks toward the sound of the katydid. Unless the katydid is frightened away, it may be found at or close to where the searchers meet or cross.

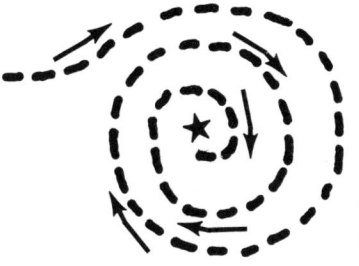

The spiral method is used when there is only one person. The searcher listens carefully, and spirals in toward the sound.

Finding the katydid.

the katydid is singing. If you are alone, try moving around the source of the song in a shrinking *spiral* that leads you to the insect.

In either case, move slowly and stop walking as soon as the katydid stops singing. When it starts again, move on slowly. When you spy the insect, spread your hands apart on either side of it. Move your hands together *slowly* until they are close to the insect. Then bring your hands quickly together, shooing the katydid into the jar you are holding in one hand.

Catching a katydid with your hands and a jar.

Don't be disappointed if you cannot catch a katydid right away. "The first time you start looking for katydids, you'll have a terrible time even *seeing* them," says Dr. Morris.

Like other scientific collectors, Dr. Morris sometimes uses a net to catch katdydids, but he doesn't recommend one for young people in most situations. He believes an amateur will do better using hands and a jar. Many of the larger katydids are slow-moving, he adds, and can even be taken by a quick movement of the hand without a jar. Hold your hand in a partial fist, and make a grab for the insect. But be careful not to squeeze it, as its jumping legs are easily detached.

Many night-singing katydids sing in trees, but luckily for amateur collectors, some stay on or near the ground. The flightless katydids common in the western United States look for food at night, so if you live in that area, you will probaly find large numbers of katydids scurrying across the ground at night. They are most easily spotted on roads. Some other night-active katydids sing in low vegetation. The same procedure can be used to catch them as with day-active katydids, but you will need a headlamp to see them.

Headlamps—a head band with a light on it—can be bought from Recreation Equipment, Inc., 1525 Eleventh and Pine, Seattle, Washington 98122, under the name "Ray-o-vac." If a headlamp is not handy, take along a friend who can carry a flashlight or lantern.

One caution: some katydids bite, a few of them hard enough to draw blood. Katydid bites are never dangerous, but they hurt. "It feels like being pinched by a pair of tweezers," says Dr. Rentz. Collectors can avoid bites, he says, by handling katydids carefully. "Cup them in the hand rather loosely, and if there's no pressure put on them, they probaby won't bite," he advises.

Once you have your katydid, what do you do with it? If you want to mount dead specimens for a collection, the necessary equipment is readly available from biological supply houses. See page 88 for a list of some of these firms. You can either buy a whole insect preparation kit or buy items separately. Here are the items you will need:

    a "killing jar"
    a mounting board or insect display case
    insect pins

A killing jar that contains ethyl acetate is recommended, as it is less dangerous to handle. A jar like this is really a small gas chamber that kills insects quickly and painlessly. Once the insect is dead, you can pin it onto a mounting board. If you want to store the insects or carry them to school, you will need a display case. A display case is simply a closed box with a mounting board lining the bottom. You can either buy one at a biological supply house or make your own, using either a cigar box or a clear plastic box. A styrofoam block can also be used to display insects.

Entomologists prepare insects before mounting so they will not decompose or lose their coloring and shape. Dr. Rentz makes a slit down the abdomen of the dead insect and removes the contents. A pair of manicure scissors can be used to make the slit. Then he dusts the empty abdomen with a mixture of three parts talcum powder and one part boric acid powder. A little loose cotton is added to keep the body shape. If you push the edges of the slit back together with your fingers, it will remain closed as the insect dries.

Green katydids prepared in this way keep most of their bright coloring indefinitely, according to Dr. Rentz. If the insect has been dead for some time before you mount it, it will be dry and you will probably have to place it in a humidity chamber.

An insect collection is fun to make, but it is even

Equipment for mounting dead specimens.

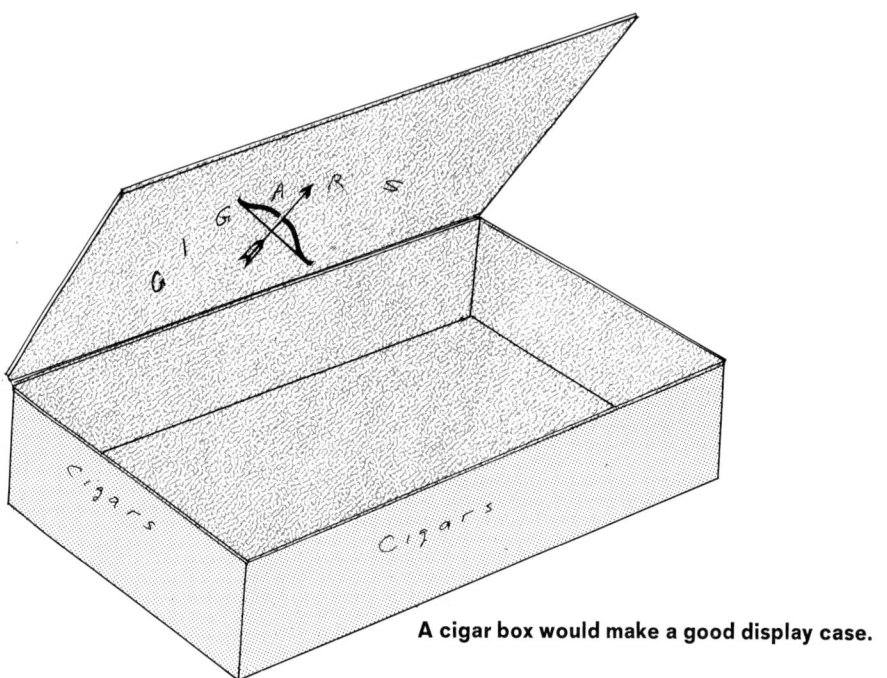

A cigar box would make a good display case.

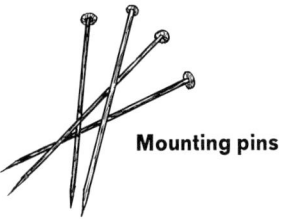
Mounting pins

Killing jar with ethyl acetate in it.

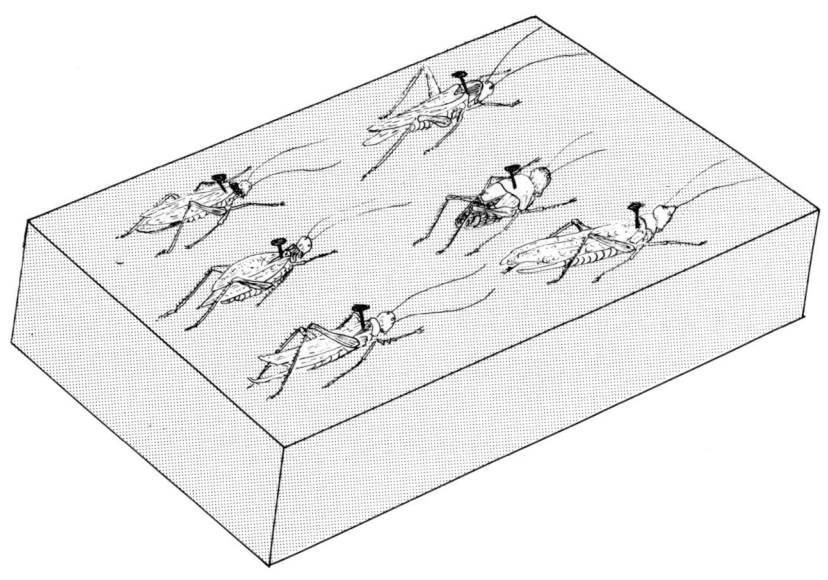

Styrofoam block with six katydids mounted on it with pins.

more fun to observe live katydids. Keep katydids in small separate containers unless you want to observe mating behavior. Otherwise, they will eat each other. A large jar with a screen top makes a good cage, and so does a screen cylinder. Don't use a jar with a lid with holes punched in the top because it is too hot for katydids.

Dr. Morris makes his own screen cylinders out of pieces of soft fiberglass screening. He rolls a piece into a cylinder shape and staples it together. Then

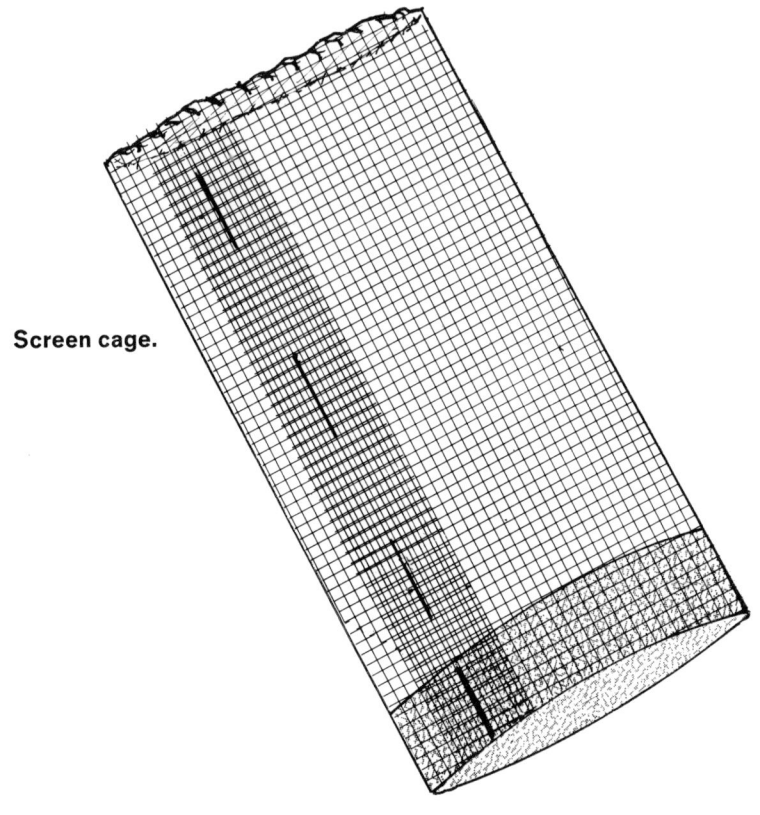

Screen cage.

he cuts out a circular piece of screen to fit the top of the cylinder and sews it to the cylinder with strong thread. He leaves the bottom open. To keep the cylinder stiff, he staples a small strip of cardboard around the bottom on the inside.

To feed your katydids, use small pieces of apple or some lettuce and a little dry dog food. First place the food next to the cage, then quickly move the cage over the food. Water the cage once a day with a mister so that water clings to the screen.

Day-active katydids can be kept on a windowsill, but night-active ones do better in a dark corner of a room. However, young katydids of any species need light to develop.

If you want to observe mating behavior, you will need a larger cage. Dr. Rentz recommends a five-gallon aquarium with a screen top or a screen cylinder at least a foot high. Place a pie pan in the bottom of the cylinder along with some earth and a branch with leaves. This will take care of the egg-laying needs of most females. Later, you may see the female lay eggs with her ovipositor.

Other behavior you can expect from your katydid are walking, jumping, grooming and, if it is a male, singing. Field recording of katydid songs requires complex equipment, but indoors you can use a tape recorder placed next to the cage to pick up songs.

By keeping a captive katydid, you can discover for yourself some of the secrets of these insects.

Biological Supply Houses

American Biological Supply Company
1330 Dillon Heights Avenue
Baltimore, Md. 21228

Ward's Natural Science Establishment, Inc.
P.O. Box 1712, Rochester, N.Y. 14603
P.O. Box 1749, Monterey, Calif. 93940

Carolina Biological Supply Company
Burlington, N.C. 27215
Powell Laboratories Division, Gladstone, Ore. 97027

Write these companies for a catalog first before you order.

Glossary

Aesop—A Greek author who lived from about 620 B.C. to 560 B.C.

antennae—the thread-like "feelers" that project from many insects' heads.

audio—a sound humans can hear.

barb—a sharp point.

behavior—the way in which an animal acts.

bog—a low area of wet, soggy ground, usually with grass but no trees.

cannibalism—eating a member of your own species.

characteristic—a feature that sets something or someone apart from others.

chorus—a group of singers who sing together.

classification—arranging animals or plants in groups by examining their characteristics.

complexity—being made up of many related parts.

entomologist—a scientist who studies insects.

fable—a story that teaches a lesson, often by using animals that speak like human beings.

fertilized—able to produce a new individual.

frequency—the number of times an object vibrates in a period of time.

humidity chamber—a box containing a sponge soaked in water, and some moth crystals to keep out mold.

insect—an animal with six legs and a three part body: head, chest or thorax, and abdomen.

killing jar—a jar containing a poison that kills insects quickly.

Linnaeus—an 18th century Swedish scientist who worked out a system of grouping animals and plants.

metamorphosis—the development of an adult by passing through a number of stages of growth.

migrate—to move from one area to another.

mimic—an animal that looks, sounds or acts like another.

molt—shedding the outer covering of an animal.

mutation—a sudden and permanent change that produces an individual unlike its parents.

nymph—a young katydid.

order—a large division of the Linnaean system.

Orthoptera—the order of insects to which the katydids belong.

orthopteran—a member of the order Orthoptera.

oscilloscope—a machine that draws a picture of a sound. The picture shows changes in loudness in a period of time.

ovipositor—a long, swordlike structure that sticks out of the rear of the female katydid's body. She uses it to lay eggs.

pesticides—chemicals used to kill insects and other pests.

prey—a creature hunted for food.

protein—a substance in many foods that we must have to live.

solitary—living alone. The opposite is social.

song—a number of sounds organized in a certain way. The sounds are repeated in the same way each time the song is given.

soundproofed—constructed so as to keep out sound.

sperm—the male cell that joins with the female egg to produce a new individual.

species—the smallest division of the Linnaean system. Members of the same species can mate and bear young which can also mate and bear young.

specimen—an individual animal or thing typical of its group.

spiral—a curve that moves around a fixed center at increasing or decreasing distances from the center.

subfamily—a group of related animals that have many characteristics in common.

temperate climates—climates with seasons which are not very hot and not very cold.

theory—a possible explanation based on evidence.

tropics—hot, wet areas near the equator.

ultrasonic—a sound humans cannot hear because the frequency is too high.

vibrate—to tremble or move rapidly back and forth.

vibration—a rapid trembling movement.

Index

A

Abert Lake, 65
Africa, 67
American cockroach, 70
American Museum of Natural
  History, The, 19
angular-winged katydids, 21,
  22, 23, 27, 37, 76
antennae, 11, 13-14, 41, 66
aquarium, 87
Arkansas, 72
Asia, 67
audio sounds, 43-44
Augusta College, 24
Augusta, Georgia, 24

B

barbs, 72
biological supply houses, 82-83
bog katydids, 41-44
Borneo, 63
Bozeman, Montana, 54
British Isles, 69
bush katydids, 37, 55, 76

C

California, 55
California Academy of Sciences,
  57
Canada, 19, 41, 67
cannibalism, 23
cellucotton, 60
Central America, 60
characteristics, 13-17
Chicago, 69
Chinese, 69
chorusing, 36-39
cigar box, 60, 61, 83
classification, 64-65
cockroaches, 13, 66, 69-70
color, 16, 66
Cornell University Laboratory
  of Ornithology, 76
Corvallis, Oregon, 65
cone-headed katydids, 17, 37
"courtship song," 44
crickets, 13, 49-56, 66, 68, 69,
  74

D

day-singing katydids, 11, 77
desert locust, 68
deserts, 57, 60
Detroit, Michigan, 22
Dew, Donald, 36
display case, 82-83
dog chow, Purina, 61

E

ears, 14, 24, 66
eggs, 16, 24, 25, 28, 60, 87
Egypt, 67
entomologists, 21, 32, 33, 54,
  69, 70, 83

ethyl acetate, 82
experiments, 42, 44, 46, 75, 76

F

Fabré, Henri, 22-23
family, 13, 16
farmers, American, 49, 53, 72
federal government, 53
fertilized egg, 24. See also eggs
field cricket, 68-69
Field Museum of Natural
  History, 69
file, 30, 36
Florida, 19, 20

G

Gangwere, Dr. Stanley K., 22
German cockroach, 70
Germany, 32
grasshoppers, 13, 21, 23, 33, 61,
  62, 66, 67
Gray, Alice, 19
Great Salt Lake, 51
Grove, Dr. Davidson, 21-22, 27

H

headlamp, 81-82
Henry, Dr. John, 54
hopper houses, 32-33
house cricket, 68-69
humidity chamber, 61, 83
Hutchins, Dr. Ross, 32

I

Idaho, 54
Indians, 49
Indonesia, 63
insect collections, 61, 63, 83
insect pins, 82
insects, 11, 13, 17, 21, 32, 36,
  39, 41, 44, 49, 50, 52, 53,.
  55, 56, 60, 61, 64, 68, 69
Ithaca, New York, 76

J

Java, 63

K

Kethley, John, 69
Kevan, Dr. D. K. McE., 33
killing jar, 60, 82

L

laboratory, 36, 40, 44, 46, 61
Lake Superior, 44
"leaf mimics," 12-13, 64
lichen mimic, 64
life cycle, 21-28, 39
Lightfoot, David C., 65
Linnean system, 13, 17
Linnaeus, 13
locusts, 66-68
longhorned grasshoppers, 13, 66
Los Angeles, 57

M

Malaysia, 63
Malibu Beach, 57
mating, 26, 85, 87
meadow grasshoppers, 37
meadow katydids, 46-48
meat eaters, 22, 72
metamorphosis, 27
migrate, 54-55, 67
mirrors, 30
molts, 26-27
Mormon crickets, 52-56, 67
Mormons, 49-51
Morris, Dr. Glenn K., 32, 40-48, 77, 81, 85
moth crystals, 61
mounting board, 82-83
music, 29
mutations, 19

N

net, 60, 81
Nevada, 57, 58, 60
New York City, 19
Nickle, David, 20
night-singing katydid, 81
nymphs, 21, 26, 27, 61

O

oblong-winged katydid, 37
Oklahoma, 72
orders, 13
Oriental cockroach, 70
Orthoptera, 13

orthopterans, 13, 66, 72
oscilloscope, 35-36, 45-46
outbreaks, 52-53, 67
ovipositor, 17, 24, 87

P

pests, 52, 55, 70
pesticides, 53, 55, 74
Philadelphia Academy of Natural Sciences, 61, 63, 64
plant eaters, 21, 22
powder, boric acid, 83; talcum, 83
praying mantis, 13, 66, 72, 74
prey, 72
protein, 56

R

Rangeland Insect Laboratory, 54
Recreation Equipment, Inc., 81
relatives, 11, 66
Reno, Nevada, 58
Rentz, Dr. David C., 57, 58, 60-61, 63-65, 77, 82, 83, 87
reproductive cells, 24
research collection, 64. See also Philadelphia Academy of Natural Sciences.
Rocky Mountain locust, 67
Rogers, Charlie, 70
round-headed katydids, 55

S

Salt Lake City, Utah, 49, 51

San Francisco, California, 57
scientific journals, 60, 65
scientists, 12, 13, 17, 21, 22, 23, 29, 32, 35, 44, 54, 72, 75
scraper, 30, 36
screen cages, 85-86
seagulls, 51-52
Seattle, Washington, 82
shield-backed katydids, 52, 55, 58, 64
shorthorned grasshoppers, 13-14, 23, 33, 66, 67
snowy tree cricket, 68-69
song, 14, 16, 23, 24, 28, 29-39, 42-46, 69, 75, 77, 87
Songs of Insects, The, 76
sound, 31, 35-36, 41-45
sound equipment, 35, 43
South Africa, 60
South America, 60, 64
South American katydid, 17
species, 17, 19, 22, 24, 39, 41, 45, 47-48, 57-58, 64-65, 67, 70, 76
specimens, 60, 62, 77, 82
sperm, 24
spiral procedure, 80
Spooner, Dr. John D., 24
styrofoam block, 83
subfamilies, 17-18, 47, 52, 58, 64

T

tape recorder, 45, 46, 87
temperate climates, 25

temperature, 39, 44, 54
Temple Square, 52
test tubes, 60, 77
Texas bush katydids, 24, 76
theory, 43
Thunder Bay, 44
tick(ing), 24, 30, 76-77
Toronto, Canada, 32, 47
triangulation, 77
true katydids, 11, 27, 37

U

Uhler's katydid, 36
ultrasonic sound, 35, 43-44, 46
U.S. Department of Agriculture, 70
United States, 11, 17, 19, 21, 52, 55, 57, 61, 67, 68, 70, 72, 76, 81
University of Florida, 20, 23, 36
University of Toronto, 32

V

vibrations, 34-35

W

Walker, Dr. Thomas J., 23, 36, 39, 75, 76
walkingsticks, 13, 66, 70
Wayne State University, 22
Westchester County, 19
wings, 12, 13, 15-17, 26, 27, 29, 30, 35, 58, 66, 68
"wrestling matches", 47-48